John Soane

# Plans, Elevations, and Sections of Buildings

executed in the counties of Norfolk, Suffolk, Yorkshire, Staffordshire,

Warwickshire, Hertfordshire, et cetera

John Soane

**Plans, Elevations, and Sections of Buildings**
*executed in the counties of Norfolk, Suffolk, Yorkshire, Staffordshire, Warwickshire, Hertfordshire, et cetera*

ISBN/EAN: 9783744680905

Printed in Europe, USA, Canada, Australia, Japan

Cover: Foto ©berggeist007 / pixelio.de

More available books at **www.hansebooks.com**

# ·PLANS· ELEVATIONS· AND ·SECTIONS·
## ·OF · BVILDINGS·

·EXECVTED · IN · THE · COVNTIES · OF·

·NORFOLK·

·SVFFOLK·

·YORKSHIRE·

·STAFFORDSHIRE·

·WARWICKSHIRE·

·HERTFORDSHIRE·

·ET·CAETERA·

## · BY · IOHN · SOANE · ARCHITECT ·

·MEMBER · OF · THE · ROYAL · ACADEMIES·

·OF · PARMA · AND · FLORENCE·

## · LONDON · MDCCLXXXVIII ·

·PVBLISHED · BY · MESS^{RS.} · TAYLOR · AT · THE · ARCHITECTVRAL · LIBRARY · · HOLBORN ·

## To the K I N G.

ENABLED by Your Majefty's Munificence to finifh my Studies in Italy, and flattered with your Permiffion for this Dedication, I am induced to hope that the fmall Tribute of a grateful Heart, will not be unfavorably received ; and that your Protection will be extended to a Work, which owes its Origin to your Patronage.

To Your Majefty's Liberality the Arts are greatly indebted, encouraged by you, they have fucceeded, and that they may long enjoy your Countenance and Support, is the unfeigned Wifh of,

Your Majefty's,

moft dutiful, and

moft faithful Subject,

J O H N   S O A N E.

Welbeck-Street,
September, 1788.

# LIST of SUBSCRIBERS.

## THE KING.

### A.

The Royal Academy
The Right Hon. Lord Arundell, Wardour-Castle
Sir Edward Astley, Bart. Norfolk
Jacob Astley, Esq.
Mr. Atkins
Mr. John Armstrong
Robert Adam, Esq.

### B.

The Marquis of Buckingham, Stowe-House
Earl of Belborough, Rowhampton
Sir Francis Basset, Bart. Tehiddy, Cornwall
John Bathoe, Esq. Bath
Samuel Bosanquet, Esq. Forest-House, Essex
Thomas Bowdler, Esq.
M. S. Branthwayt, Esq. Taverham, Norfolk
William Beckford, Esq. Fonthill
Rowland Burdon, Esq. Castle-Eden, Durham
Mr. Bayley
Mr. T. Baldwin, Bath
Mr. Thomas Bradley, Halifax, Yorkshire
Mr. G. Byfield
Nat. Bassnet, Esq.
Mr. John Bevan
Mr. William Blandy, Reading, Berks
Mr. Samuel Baker, Rochester
Henry Bell, Esq. Wallington, Norfolk

### C.

The Right Honorable Lord Camelford, Petersham, Surry
Sir William Chambers, Whitton, Middlesex
Lieutenant General Cowper, Ham, Surrey
Charles Collyer, Esq. Gunthorpe, Norfolk
Mr. Cornish, Exeter
William Colhoun, Esq. M. P. Norfolk
James Crowe, Esq. Jack's Wood, Norfolk
Charles Coggan, Esq.
Captain John Coggan
Rev. Wm. Coxe, F. R. S. Bemerton, near Salisbury
Mr. Cheney, Nantwich, Cheshire
Mr. Clark
Mr. Cantwell
Mr. William Colcot
Mr. Robert Chapman

### D.

The Right Honorable Lord Dover, Rowhampton
B. G. Dillingham, Esq. Letton-Hall, Norfolk
John Denniston, Esq. Ossington, near Newark
John Richard Dashwood, Esq. Cockley-Cley, Norfolk
T. Dowdeswell, Esq. Worcestershire
Edward Darell, Esq. Richmond, Surry
Mr. Dickenson
Mr. Thomas Dove, Norwich
Mr. De Carle, Norwich

### E.

Mr. Ewen, Norwich
Mr. Thomas Egerton

### F.

The Right Honorable Lord Fortescue, Castle-Hill, Devonshire
The Right Honorable Lady Fortescue
Robert Fellowes, Esq. Shottisham, Norfolk
Sir William Fordyce
Mr. Foxhall
Mr. John Fuller, jun. Liverpool
Mr. John Fulcher, Bury, Suffolk
Mr. Thomas Fulcher, Ipswich

### G.

Mr. Goslet, Viscount of Jersey
Payne Galway, Esq. Tofts, Norfolk
Thomas Giffard, Esq. Chillington, Staffordshire
The Reverend Archdeacon Gooch, Saxlingham, Norfolk
G. de Ligne Gregory, Esq.
Mr. Goudge
Mr. Robert Golden
Mr. William Goodwin
Mr. Geo. Griffiths

### H.

The Earl of Hardwicke, Wimple
The Right Honorable Lord Herbert
The Honorable John James Hamilton, Bentley-Priory, Middlesex
Sir Alexander Hood, K. B. Cricket-Lodge, Somersetshire
Henry Holland, Esq. 2 sets
Mr. R. Holland
Rev. Gervas Holmes, Gawdy-Hall, Suffolk
William Herring, Esq. Norwich
Robert Hunter, Esq. Thurston by Dunbar
James Hunter, Esq.
James Hatch, Esq. Oldford
Thomas Hammersley, Esq.
Mr. Hakewill
Mr. Holroyd
Mr. Handasyde
Mr. John Hyram
Mr. D. Harris, Oxford
Mr. William Hobson
John Hele, Esq.
Mr. Hobcraft
Richard Hill, Esq.

### I.

Sir John Ingoldby, Bart.
John Johnson, Esq.

Robert

# LIST of SUBSCRIBERS.

Robert Jones, Efq. Fonman-Caſtle, Glamorganſhire
Thomas Johnes, Efq. M. P. Croft-Caſtle, Herefordſhire
Mr. G. Jernegan

## K.

The Right Honorable Lord Kinnaird

## L.

Sir James Tylney Long, Bart. Wanſtead-Houſe
H. G. Lewis, Efq. Malvern-Hall, Warwickſhire
James Lewis, Efq
Thomas Latter, Efq. Gadebridge-Houfe, Herts
Samuel Lapidge, Efq. Hampton-Court
John Larking, Efq. Eaſt-Malling, Kent
Thomas Leverton, Efq.
Mr. Lewington, Reading
Mr. William Lumley
Mr. William Lovering
Mr. Richard Long, Sudbury, Suffolk

## M.

The Right Honorable Lord Mulgrave, Mulgrave-Hall, Yorkſhire, a ſets
The Right Honorable Lord Macartney
Sir John Morſhead, Bart. M. P.
General Murray, Battle, Suſſex
Richard Milles, Efq. Elmham, Norfolk
William Morland, Efq.
Jeremiah Milles, Efq. Piſhiobury, Herts
Robert Marſham, Efq. F. R. S. Norfolk
Nathaniel Middleton, Efq.
Thomas Morton, Efq.
Mr. Marks, Norwich
Mr. Moulton

## N.

The Worſhipful Corporation of the City of Norwich
The Duke of Northumberland, Sion-Houfe
George Nelthorpe, Efq. Lyndford-Hall
Mr. Nelſon
Mr. Neill

## O.

The Earl of Orford, Houghton, Norfolk
James Oakes, Efq. Bury, Suffolk
William Ord, Efq. Fenham North

## P.

The Duke of Portland, Bulſtrode-Park
The Right Honorable Lord Petre, Thorndon-Houſe
The Right Honorable William Pitt, Hollwood, Kent
Sir Thomas Beauchamp Proctor, Bart. Langley-Park, Norfolk
John Parteſon, Efq. Norwich
Edward Roger Pratt, Efq. Ryſton-Hall, Norfolk
Sir Robert Palk, Bart. Haldon-Houſe, Devonſhire
James Peacock, Efq.

Richard Page, Efq. Wembley, Middleſex
John Peachey, Efq. Weſt-Dean, Suſſex
Mr. John Pentland, Dublin
Mr. John Painter, Emſworth, Hampſhire
Mr. John Papworth

## R.

The Earl of Roſeberry, Bixley-Hall, Norfolk
Sir Joſhua Rowley, Bart. Tendring-Hall, Suffolk
Lady Rowley
Mr. Thomas Rathbone, Cheſter
Mr. George Richardſon

## S.

The Right Honorable Lord Sondes, Rockingham, Northamptonſhire
The Right Honorable Lord Southampton, Highgate
George Smith, Efq. Piercefield, Monmouthſhire
Joſhua Smith, Efq.
Drummond Smith, Efq.
John Stockwell, Efq.
Mr. Scrimpſhaw
Mr. Shepherd
Thomas Saunders, Efq.
Mr. Joſeph Sibley

## T.

The Honorable Wilbraham Tollemache, Steep-Hill, Iſle of Wight
John Throckmorton, Efq. Weſton, Bucks
Mr. Tyrrell
Mr. Turtle
Mr. S. Townſend, Oxford

## W.

Sir Bourchier Wrey, Bart. Tawſtock. Devon
The Honorable Lewis Thomas Watſon, Lees-Court, Kent
Sir George Warren, Bart.
The Reverend Philip Wodehouſe
William Windham, Efq. Earſham, Suffolk
Joſeph Windham, Efq.
Mr. George Wyatt
Nehemiah Winter, Efq.
Ralph Winter, Efq. Hockerill, Eſſex
John White, Efq.
Thomas Wildman, Efq.
Nathaniel Wright, Efq.
The Rev. John Wheler
Mrs. E. Wheler
John Wharton, Efq. Skelton-Caſtle, Yorkſhire
Mr. Watſon
Mr. Witherby

## Y.

Philip Yorke, Efq. M. P. Hammels, Herts
Lady Elizabeth Yorke
John Yorke, Efq. Richmond, Yorkſhire
Charles Yorke, Efq.

# INTRODUCTION.

It will be needless to apologize for the following observations, since custom has so fully established the propriety of an introductory address from all who present their labours to the public.

ARCHITECTURE, the subject of the present work, no less delightful in itself, than calculated to increase the comforts and conveniencies of mankind, was anciently held in the highest estimation. Not only its patrons, but its professors, were in the first class of men; and every wise and great prince has always had recourse to architecture to perpetuate his name. In ancient times it was great and meritorious to raise the temple, the portico, and other public edifice. How great the advantage and glory that accrued to the Roman name and empire from their buildings, the amphitheatres, triumphal arches, baths, aqueducts and other remains of ancient magnificence abundantly testify. The monuments and trophies that were raised at the public expence to perpetuate the memory of great atchievements, at the same time that they immortalized the fame of individuals, were lasting proofs of the justice and liberality of the people; they stimulated others to engage in the service of their country, to exert themselves in honourable actions, and strongly induce us to believe many things recorded by their historians, which might otherwise have been deemed incredible.

QUANTA autorità habbia arrecato lo edificare allo imperio et nome Romano, non accrescerò io con il mio dire, più che quella che noi per i sepolchri & per le reliquie dell' antica magnificentia, sparse per tutto, veggiamo haverne data cagione che si presti fede, a molte cose dette dalli historiografi, le quali forse altrimente sarebbono parute incredibili —— et chi è stato quello infra i grandissimi & prudentissimi

B

prudentiſſimi principi, chi tra le prime lor cure, ò penſieri di perpetuare il nome, et la poſterita ſua, non ſi ſia ſervito della architettura?

L. B. Alberti il Proemio.

Vitruvius informs us, that it was the cuſtom of the ancient artiſts to commit their inventions and improvements to writing, and has left us a large liſt of authors whoſe works the devouring hand of time, and the fury of barbariſm have deſtroyed. With what heartfelt regret muſt every man of genius reflect on the loſs of theſe numerous treatiſes, compoſed by men whoſe ambition was to elevate the ſcience, and to inſpire the riſing artiſts with the ſame enthuſiaſm which they felt !

Majores eum ſapienter, tum etiam utiliter inſtituerunt per commentariorum relationes cogitata tradere poſteris.

Vitr. Lib. vii. Præf.

Vitruvius is the only ancient author on architecture now extant, and from him much may undoubtedly be collected. He is the father of architects, and writes with the zeal of a man anxious to raiſe his profeſſion; he has enlarged on the qualifications neceſſary to form a great artiſt ; has placed the art in the moſt honourable point of view ; and rendered the profeſſors reſpectable ; he particularly inculcates the neceſſity of philoſophy to enlarge the mind of the artiſt, to free him from arrogance, and to make him courteous, juſt and faithful ; above all things he exhorts him to avoid avarice ; as no work can ſucceed without fidelity and integrity; and not to be covetous, nor to have his mind intent on receiving gifts, but to ſupport with prudence a.. ' propriety, his dignity and reputation.

Philosophia vero perficit architectum animo magno, et uti non ſit arrogans ſed potius facilis, æquus et fidelis, ſine avaritia, quod eſt maximum, nullum enim opus vere ſine fide et caſtitate fieri poteſt : ne ſit cupidus, neque in muneribus accipiendis habeat animum occupatum, ſed cum gravitate ſuam tueatur dignitatem bonam famam habendo.

Vitr. Lib. i. Cap. 1.

Sed forte nonnulli hæc levia judicantes, putant eos eſſe tantum ſapientes, qui pecunia ſunt copioſi. Itaque plerique ad id propoſitum contendentes audacia adhibita cum divitiis etiam notitiam ſunt conſecuti.

Vitr. Lib. vi. Præfatio.

Neque

Nequi eſt mirandum quid ita pluribus ſim ignotus. Cæteri architecti rogant et ambiunt, ut architectentur: mihi autem a præceptoribus eſt traditum, rogatum non rogantem oportere ſuſcipere curam, quod ingenuus color movetur pudore, petendo rem ſuſpicioſam, nam beneficium dantes, non accipientes, ambiuntur.

VITR. Lib. vi. Præfatio.

MAJORES primum a genere probatis opera tradebant architectis, deinde quærebant, ſi honeſte eſſent educati: ingenuo pudori, non audaciæ protervitatis committendum judicantes.

VITR. Lib. vi. Præfatio.

L. B. ALBERTI, who wrote exprefsly to rival Vitruvius, though he has failed in the attempt, has notwithſtanding left many uſeful precepts. Faithful to the text and doctrine of Vitruvius, he has joined him with equal zeal, in defining what an architect ſhould be: I will not, ſays he, rank the mechanic with the architect; but I ſhall call him an architect, who, from his earlieſt youth, by long and extenſive ſtudy, has acquired abilities to deſign, and judgment to execute great and uſeful works, only to be effected by men of ſcience.

Io non ſi porrò innanzi un legnajuolo che tu lo habbi ad aguagliare ad huomini nelle altre ſcienzie eſſercitatiſſimi. Architettore chiamerò io colui il quale ſaprà con certa et maraviglioſa ragione et regola, ſi con la mente et con lo animo diviſare, ſi con la opera recare a fine tutte quelle coſe, &c.

ALBERTI il Proemio.

THE Grecian artiſts travelled into Egypt in order to enrich their minds with uſeful knowledge; and the Romans, in ſucceeding ages, ſought perfection in Greece, hoping to rival in the arts, thoſe whom they had conquered by their arms; the modern artiſts, treading in the ſame path, viſit Italy to correct their taſte, and to enlarge their ideas. The great remains of antiquity exhibit many glorious examples of the ſublimity of the arts, and the perfection of ancient ſculpture and architecture; affording us ſome conſolation for the loſs of the many invaluable treatiſes of the ancients, mentioned by Vitruvius and others. But as every man was not an Apollodorus, a Dinocrates, or an Hermogenes, let us not therefore blindly and ſervilely copy the ancient buildings, but cautiouſly examine them, and if poſſible catch the ſpirit of them: by conſtant ſtudy, deep reflection, and unwearied diligence, we ſhall diſcover the cauſes of their various

combinations

combinations and proportions, and shall trace the springs from whence we derive satisfaction in contemplating the venerable remains of ancient grandeur; we shall then look upon those wonderful and stupendous works, with equal pleasure and improvement; we shall constantly discover new beauties; we shall perceive how different are the effects produced by the light of the objects themselves, from the ideas raised on examining them in prints, drawings and models; we shall see how closely the ancient artists attended to the character, convenience and locality of their edifices; and that the same ornaments, and the same proportions that astonish and delight in some situations, fail of effect in others.

Nec tamen in omnibus (operibus) symmetriæ ad omnes rationes et effectus possunt respondere, sed oportet architectum animadvertere, quibus proportionibus necesse sit sequi symmetriam, et quibus rationibus ad loci naturam aut magnitudinem opus debeat temperari.——Si qua alia intercurrunt, ex quibus necessitas cogit discedere ab symmetria, ne impediatur usus.——Hoc autem erit, si architectus erit usu peritus, præterea ingenio mobili solertiaque non fuerit viduatus.

Vitr. Lib. v. Cap. 7.

The great masters of the fifteenth and sixteenth centuries were indefatigable in their researches into the monuments of antiquity, uniting in their studies painting, sculpture and architecture; together with the most extensive knowledge of the various sciences depending on those arts. Their numerous works in Rome, and other parts of Italy, point out to us the happy effects of this union. Let us examine, therefore, the works of Raphael, Michael Angelo, Julio Romano, Palladio, Scamozzi, Vignola, and the other great restorers of architecture, and studiously observe, how cautiously they used the inestimable remains of antiquity. From their labours, and the study of the ancient buildings, we learn the necessity of long, extensive, and close application, and the impossibility that any man should arrive at a tolerable knowledge and perfection in architecture, without having been previously trained to the arts from his earliest infancy, and nursed, as it were, in the bosom of science.

Cum ergo tanta hæc disciplina sit condecorata et abundans eruditionibus variis ac pluribus, non puto posse juste repente se profiteri architectos, nisi qui ab ætate puerili his gradibus, disciplinarum scandendo scientia plurium literarum et artium nutriti, pervenerint ad summum templum architecturæ.

Vitr. Lib. i. Cap. 1.

The

THE ancient artifts, and the great reftorers of architecture attained the fummit of reputation, fame, and profit, by flow and gradual advances; but enterprifing and interefted mechanics, more anxious to acquire wealth, than to fecure fame, have found fhorter and eafier roads to fuccefs, though not to fcience, and by following the precept of Martial,

> Si duri puer ingenú videtur,
> Præconem facias, vel architectum.
>
> MART. Lib. v. Epigr. 56.

have prodigioufly encreafed the number of architects, and furveyors. In the prefent times there is a fafhion even in architecture; a fafhion which renders learning and application needlefs, and teaches men boldly to attempt every thing; a fafhion —— that has brought forward men, whofe works replete with foreign abfurdities, future ages will view with wonder and aftonifhment. Doubtlefs the judicious artift will find many things in the arrangement and decorations of modern French and Italian boufes worthy of his ferious attention; but the abfurdities daily intruded on us for French refinements, introduced without the leaft regard to difference of climate, and mode of living, are too grofs to efcape cenfure.

ARCHITECTURE is a coy miftrefs that can only be won by unwearied affiduities, and conftant attention; but when the mind is wedded to it, the imagination is always filled with wonder and delight, and the poffeffor feels himfelf well rewarded for the trouble of purfuit; indeed fo fafcinating is the ftudy of architecture that many men with fortune and talents have devoted their time to the attainment of a fcientific knowledge of its principles, and few have the means without the inclination for building; many of the comforts of life are heightened by the conveniences of our manfions; we look with pleafure on each man's improvements, and feel real fatisfaction at the fight of every well-contrived and ingenious defign, where beauty, elegance and convenience unite.

C

IN

In building it is of great confequence not to begin haftily, for the defects of a work are often feen and felt when the beauties are unnoticed and forgotten. The greateft exertion of judgment, experience and attention is requifite in compofing defigns; that we may not be led away with a vain defire of introducing novelty and paltry conceits at the expence of propriety and convenience. Variety in the compartitions, eafy communications, and well-placed ftair-cafes, each part entire in itfelf, and all tending as rays to a center, are neceffary to produce a convenient, elegant and harmonious whole, that may engage the attention, and fecure the praife of the judicious; while hafty and imperfect productions not only occafion continual alterations,

Diruit, ædificat, mutat quadrata rotundis.

Hor. Epiftolæ, Lib. i. E. 1.

but entail lafting difgrace on their authors.

Persons of no fkill will often point out an excellence or defect in the form and deftination of a building, and in the arrangement of its parts, and may make obfervations worthy of attention; for the eye readily difcovers whatever is convenient, elegant and graceful. Let him therefore who intends to build take the opinion of his friends, as well as of profeffional men; he may then reafonably hope, to have his doubts and difficulties removed, and to poffefs all the information that nature, genius, experience and judgment can fuggeft.

Having determined to build we muft firft attend to the fituation, next to the defign of the edifice, and to the nature and quality of the materials, laftly, to a minute and particular defcription of the various works, with a correct inveftigation of the expence.

The fituation muft be carefully attended to; good water and a dry fertile foil are indifpenfable requifites, which muft not be overlooked or facrificed to beautiful fcenery, or any other confideration whatever, as nothing can compenfate for the want of thefe advantages.

The drawings being completed, a plain model of the whole building fhould be made of a fufficient magnitude to fhew the feveral parts of each floor, free from all colouring, which only deceives the eye, and diverts the attention from fcrutinizing the component parts: the fituation, forms, and connections of the feveral apartments may then be diftinctly viewed; and that a correct judgment may be formed of their proportions, examine rooms of fimilar dimenfions, particularly noticing the fituation of the doors, windows and chimneys.

Having

HAVING made the defigns as perfect as poffible, and in every refpect fully fatisfactory, yet we ought not haftily to pull down the old manfion; or lay the foundations of the new one; but take Pliny's advice on another occafion, and lay the whole entirely afide, until it ceafes to be familiar to the mind.

> POTERIS et quæ dixeris poft oblivionem retractare, multa retinere, plura tranfire, alia interfcribere, alia refcribere.
>
> PLIN. Lib. vii. E, 9.

IF on re-examination, the whole ftill appears clear and fatisfactory; full and particular defcriptions of all the different works fhould be made with the utmoft precifion and accuracy, and the earth fhould be bored in various places, and wells funk to afcertain the quality of the water, the nature of the foil, and the precautions neceffary to be taken in the foundations.

ESTIMATES are next to be confidered, which if the works are entirely new, may be made with the utmoft accuracy and certainty, whatever builders may urge to the contrary; when they are not fo, it arifes from the fame perfon being the architect, the builder, and, as is fometimes the cafe, the contractor alfo; from ignorance, or the cruel maxim of holding out fpecious inducements to begin building; well knowing that every nerve will be ftrained to avoid the difgrace and inconvenience of leaving the work unfinifhed.

THE bufinefs of the architect is to make the defigns and eftimates, to direct the works and to meafure and value the different parts; he is the intermediate agent between the employer,, whofe honour and intereft he is to ftudy, and the mechanic, whofe rights he is to defend. His fituation implies great truft; he is refponfible for the miftakes, negligences, and ignorances of thofe he employs; and above all, he is to take care that the workmen's bills do not exceed his own eftimates. If thefe are the duties of an architect, with what propriety can his fituation and that of the builder, or the contractor be united?

VITRUVIUS is particularly copious on this head, and fpeaks the language of a man preferring honour and probity, to intereft and gain; his words fhould be treafured up in the mind, and carefully adhered to by every man anxious to fupport the refpect due to his profeffion.

NOBILI

Nobili Graecorum et ampla civitate Ephesi lex vetusta dicitur a majoribus dura conditione, sed jure esse non iniquo constituta; nam architectus cum publicum opus curandum recipit, pollicetur quanto sumptu id futurum, tradita aestimatione, magistratui bona ejus obligantur, donec opus sit perfectum. Eo autem absoluto, cum ad dictum impensa respondet, decretis et honoribus ornatur: item si non amplius quam quarta in opere consumitur, ad aestimationem est adjicienda, et de publico praeflatur, neque ulla poena tenetur: cum vero amplius quam quarta in opere consumitur, ex ejus bonis ad perficiendum pecunia exigitur. Utinam Dii immortales fecissent, quod ea lex etiam populo Romano, non modo publicis, sed etiam privatis aedificiis esset constituta! namque non sine poena grassarentur imperiti, sed qui summa doctrinarum subtilitate essent prudentes, sine dubitatione profiterentur architecturam, neque patres familiarum inducerentur ad infinitas sumptuum profusiones, et ut ex bonis ejicerentur: ipsique architecti, poenae timore coacti diligentius modum impensarum ratiocinantes explicarent, uti patres familiarum ad id, quod praeparavissent, seu paulo amplius adjicientes, aedificia expedirent. Nam qui quadringenta ad opus possunt parare, si adjiciant centum habendo spem perfectionis, delectationibus tenentur: qui autem adjectione dimidia, aut ampliore sumptu onerantur, amissa spe, et impensa abjecta, fractis rebus et animis, desistere coguntur.

Vitr. Lib. x. Praefatio.

Ornaments are to be cautiously introduced; those ought only to be used that are simple, applicable and characteristic of their situations: they must be designed with regularity and be perfectly distinct in their outlines; the Doric members must not be mixed with the Ionic, nor the Ionic with the Corinthian, but such ornaments only should be used, as tend to shew the destination of the edifice, as assist in determining its character, and for the choice of which the architect can assign satisfactory reasons.

Multa ornamenta saepe in operibus architecti designant de quibus argumentis rationem, cur fecerint, quaerentibus reddere debent.

Vitr. Lib. i. Cap. i.

The

The ancients with great propriety decorated their temples and altars with the fculls of victims, rams heads and other ornaments peculiar to their religious ceremonies; but when the fame ornaments are introduced in the decoration of English houfes, they become puerile and difgufting.

After the authors and works already mentioned it would be as ufelefs as prefumptuous to enter into any detail relating to the elements and orders of architecture; the lovers of the arts will confult with pleafure and profit the parallel of the ancient architecture with the modern, written in French by Roland Freart and tranflated by Evelyn, a work of great learning and merit.

The ingenuity of mankind has hitherto produced only three diftinct orders of architecture, and perhaps never will invent more, unlefs fuch attempts as are fhewn in " A Propofition for a " New Order of Architecture" can be confidered as increafing the number; yet the Gothic architecture being entirely diftinct in all its parts from the Grecian orders gives us fome reafon to hope.

By Gothic architecture I do not mean thofe barbarous jumbles of undefined forms in modern imitations of Gothic architecture: but the light and elegant examples in many of our cathedrals, churches, and other public buildings, which are fo well calculated to excite folemn, ferious and contemplative ideas, that it is almoft impoffible to enter fuch edifices without feeling the deepeft awe and reverence. King's College Chapel at Cambridge, is a glorious example of the wonderful perfection of Gothic architecture; there is a boldnefs and mathematical knowledge peculiar to this edifice, which claims our earneft attention and admiration, which excites us to the purfuit of geometrical knowledge, and reminds us of the high opinion the ancients had of geometry.

Aristippus philofophus Socraticus, naufragio cum ejectus ad Rhodienfium litus animadvertiffet geometrica fchemata defcripta, exclamaviffe ad comites, ita dicitur, bene fperemus, hominum enim veftigia video.

Vitr. Lib. vi. Præfatio.

In this country are the moft and beft examples of Gothic architecture, in its various ftages of rife, progrefs and decline; it is therefore to be hoped fome ingenious artift will find a patron of fufficient tafte and fortune to employ his talents and preferve from deftruction, by accurate drawings and models, the mouldering remains of Gothic genius and grandeur.

D

I have

I HAVE freely borrowed from the writings of Vitruvius, L. B. Alberti, Pliny the Conful and others, therein following the example of the former:

> Ego vero (Cæfar) neque alienis indicibus mutatis, interpofito nomine meo id profero corpus, neque ullius cogitata vituperans, inftitui ex eo me approbare: fed omnibus fcriptoribus infinitus ago gratias, quod egregiis ingeniorum folertiis ex ævo collocatis, abundantes aliis alio genere copias præparaverunt, unde nos uti fontibus haurientes aquam, et ad propria propofita traducentes, fæcundiores et expeditiores habemus ad fcribendum facultates, talibufque confidentes auctoribus, audemus inftitutiones novas comparare.
>
> VITR. Lib. vii. Præfatio.

THE text of Vitruvius fhews his modefty and candor, and at the fame time furnifhes a bright example of imitation for modern artifts, but this like many of his precepts has been entirely neglected, as a late publication too plainly evinces.

> Vide WALPOLE's ANECDOTES, Vol. iv. P. 243.

IDEAL defigns have been treated, by an ingenious author, with great contempt: certainly thofe that have been executed are more to be relied on, as they muft have been better confidered and digefted, for without practical knowledge theory is of little worth, the artift converfant in the practice of building, muft have often met with difficulties after he had made drawings of every part, and attentively confidered the whole defign.

IT is impoffible to compofe one defign adapted to every fituation, an eminence and a valley require a different ftile of architecture; an edifice in an open country fhould confift of large and fimple parts, while the peaceful valley, and filent ftream admit of more delicacy and ornament. The difference in manner of living, and the different ideas of convenience, comfort

and

and elegance, render the attempt at forming one plan for every fituation ftill more impracticable.

In compofing the following defigns I have been more anxious to produce utility in the plans than to difplay expenfive architecture in the elevations; the leading objects were to unite convenience and comfort in the interior diftributions, and fimplicity and uniformity in the exterior; to collect together fome defigns of houfes and other buildings already executed, in which attention has been paid to the locality, to the different ideas of comfort and convenience, and to the ftile of living of the feveral poffeffors. If the public fhould judge as favourably of them as the individuals for whom they have been executed, I fhall flatter myfelf that my time has not been mifapplied, nor my endeavours ufelefs.

JOHN SOANE.

Welbeck-Street, Cavendifh-Square,
September 10, 1788.

# TABLE of CONTENTS.

TABLE of CONTENTS.

# · S H O T T I S H A M ·

## ·THE·SEAT·OF·ROBERT·FELLOWES·ESQ·
## ·NEAR·NORWICH·

This houfe forms half the letter H. and is fronted with white bricks of the beft quality; the fteps, window dreffings, cornices, &c. are chiefly of Portland ftone, and the capitals to the pilafters are of Coade's manufactory. The principal floor is raifed about two feet and an half.

## ·P  L  A  T  E · I ·

### ·THE·PLAN·OF·THE·PRINCIPAL·STORY·AS·EXECVTED·AND·THE
### ·ENTRANCE·FRONT·AS·INTENDED·

By four fteps you afcend the veftibule, on the right of which is the eating-room, and on the left the withdrawing-room; a fmall cabinet communicates with the withdrawing-room and library; beyond the library is a juftice-room, &c.; the beft ftair-cafe is placed in the center of the houfe, and lighted with a large Venetian window; the common ftair-cafe adjoins the offices.

   a. Leads to the waiting-room.
   b. Cabinet.
   c. Lobby to water-clofet, over which is another, with a communication from the great ftair-cafe.
   d. Paffage, &c.

## ·P  L  A  T  E    II ·

### · THE · ENTRANCE · FRONT · AS · EXECVTED ·

## ·P  L  A  T  E    III ·

### ·THE·PLAN·OF·THE·PRINCIPAL·STORY·AND·ELEVATION·OF·THE·
### ·ENTRANCE·FRONT·AS·ORIGINALLY·PROPOSED·

The plan of the manfion-houfe, in this defign alfo, forms half the letter H.

   a. Great ftair-cafe.
   b. Common ftair-cafe.
   c. c. Store-clofets.
   d. Meal-room.
   e. e. Arcade and paffage from the houfe to the offices and kitchen-court.
   f. Communication from the offices to the eating-room.
   g. Salting-bins, one placed above the other.
   All the other rooms and communications are explained in the plate.

*Entrance front, as intended*

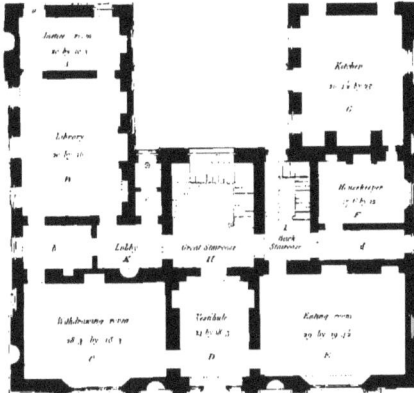

*Plan of the Principal Story*

Entrance front

Plate 5

ST OTTERHAM.

*Entrance front as originally proposed.*

*Plan of the Principal Floor.          according to the first design.*

# · M A L V E R N · H A L L ·

## · THE · SEAT · OF · HENRY · GRESWOLD · LEWIS · ESQ ·
## · NEAR · SOLYHVLL · WARWICKSHIRE ·

THE fituation of this houfe is in the middle of a park, commanding many very pleafant prof-pects; it is built with bricks, and intended to be ftuccoed; the plinths, cornices, fteps and portico are all of ftone.

The dark teints fhew the old building.

The light teints fhew the improvements.

## · P L A T E · IV ·

### · THE · PLAN · OF · THE · PRINCIPAL · STORY · WITH · THE · ALTERATIONS ·
### · AND · IMPROVEMENTS ·

A CIRCULAR portico of the Ionic order leads to the veftibule, from whence a double ftair-cafe is feen through three arches; on one fide of the hall are two drawing-rooms, a chamber, dreffing-room and ftair-cafe; and on the other are two eating-rooms, and alfo a chamber, dreffing-room and ftair-cafe; from this ftair-cafe the dinner is ferved into either of the eating-rooms, and it alfo makes the communication from the offices, to the chamber and dreffing-room, as does the ftair-cafe in the other wing, to its correfponding chamber and dreffing-room.

    a. Dreffing-room.

    b. Water-clofet.

    c. Dreffing-room.

As the dimenfions of the firft drawing-room F. were thought fufficient, the great room E. intended for a drawing-room, is finifhed as a green-houfe.

## · P L A T E · V ·

### · THE · PLAN · OF · THE · BASEMENT · STORY ·

A. A. Hot and cold baths and dreffing-rooms.

B. Store-room for foap, candles, &c.

C. C. Plate-clofet and working-room.

D. Butler's room, with a communication with the room C. by the fide of the chimney.

a. Is for the fervants to drefs in.

## · P L A T E · VI ·

### · THE · PERSPECTIVE · VIEW · OF · THE · ALTERATIONS · AND ·
### · IMPROVEMENTS ·

MALVERN HALL, WARWICKSHIRE.

Plan of the Principal Floor

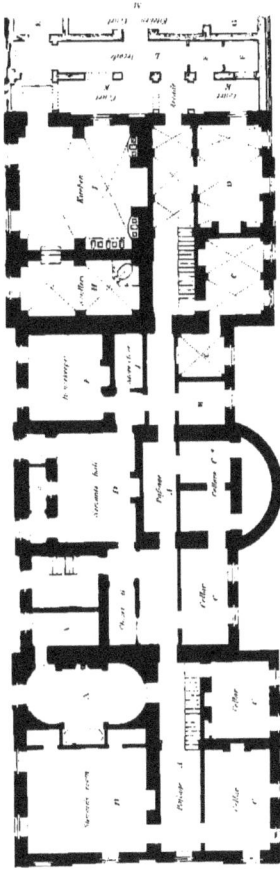

EASTERN PLAN

Plan of the Basement Story.

# · L E T T O N · H A L L ·

## ·THE · SEAT · OF · B · G · DILLINGHAM · ESQ · NEAR · SHIPDAM · IN · NORFOLK ·

THE principal ſtory of this houſe is elevated about four feet; the fronts are of white bricks, and the ſteps, columns, cornices, and other decorations are of Portland ſtone.

## · P L A T E · VII ·

### ·THE · PLAN · OF · THE · PRINCIPAL · STORY · AND · THE · ELEVATION · OF · THE · ENTRANCE · FRONT ·

A FLIGHT of ſtone ſteps leads to the veſtibule, on the right of which is a library, opening into the with-drawing-room, to which the eating-room adjoins; the breakfaſt-room is in the entrance front on the left ſide of the hall, and all the rooms have ſeparate communications; the beſt ſtair-caſe is placed in the center of the houſe, and is of Portland ſtone; and likewiſe the common ſtair-caſe.

a. Cloſet, arched and ſecured from fire for papers, records, &c.
b. A ſmall room for the butler's uſe: the offices being on the baſement ſtory, this room was fitted up with preſſes, ſink, &c.

## · P L A T E · VIII ·

### · THE · PLAN · OF · THE · BASEMENT · STORY · AND · THE · ENTRANCE · FRONT · AS · INTENDED ·

IT was propoſed to arch the whole of this ſtory, but the idea was changed after the foundations were laid, and the wine cellar only is arched.

a. b. c. d. Cellars.
e. Lobby leading to kitchen, houſekeeper's room, &c.
f. Paſſage to ſervants hall.
g. Common ſtair-caſe.
h. Room for cleaning ſhoes and knives, and for the ſervants to dreſs in.
The other rooms are particulariſed in the plan.

## · P L A T E · IX ·

### · THE · PLAN · OF · THE · CHAMBER · AND · ATTIC · STORIES ·

THE chamber ſtory contains the lady's dreſſing-room and four chambers, with dreſſing-room, cloſets, &c.
BETWEEN the principal floor and the chamber ſtory is a mezzanine (under f. g.) containing a water-cloſet, houſemaid's cloſet, a leaded ſink and the water laid on.
THE attic ſtory contains the nurſery, four chambers, two dreſſing-rooms, &c.

a. Common powdering-room.
b. c. e. Cloſets.
d. Sky-light over beſt ſtair-caſe.

## · P L A T E · X ·

### · THE · PLAN · OF · THE · PRINCIPAL · STORY · AND · ELEVATION · OF · THE · ENTRANCE · FRONT · AS · PROPOSED ·

## · P L A T E · XI ·

### · THE · PLAN · AND · ELEVATIONS · OF · THE · STABLE · BVILDINGS · AS · PROPOSED ·

THE ſtables and coach-houſes are built on a plan forming three ſides of a quadrangle, one ſide making part of the wall of the kitchen garden, and the dung is placed in a ſmall incloſed court, immediately com-municating with the garden.

a. Harneſs, ſaddle-rooms, &c.
b. Coach-houſes.

ς

LETTON HALL, NORFOLK.

*Entrance front*

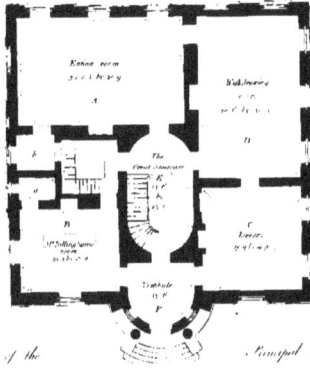

*Plan of the* ... *Principal Story*

Plate 8.

LETTON HALL

*Entrance front as intended.*

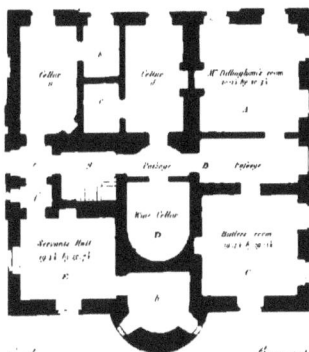

| Cellar | Cellar | Mr Dillingham's room |
| Servants Hall | Wine Cellar | Butlers room |

*Plan of the Basement Story*

Published ... by ... London.

Plate 9

GRITTON HALL

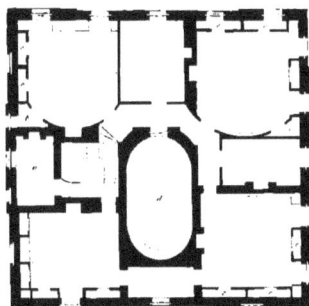

*Plan of the Attic Story*

*Plan of the Chamber Story*

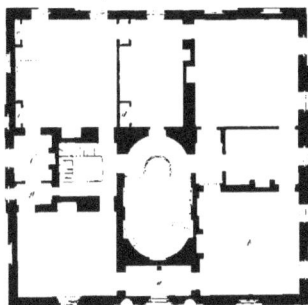

Plate IV

LETTON HALL

*Entrance front.* *as originally proposed.*

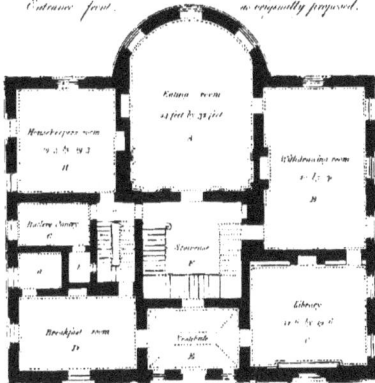

*of the Principal floor.* *according to the first design.*

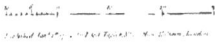

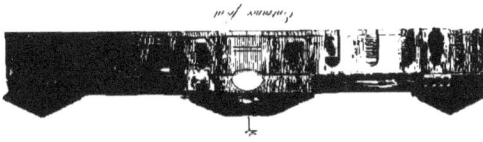

*Front of Building*

Stables at LEETON HALL.

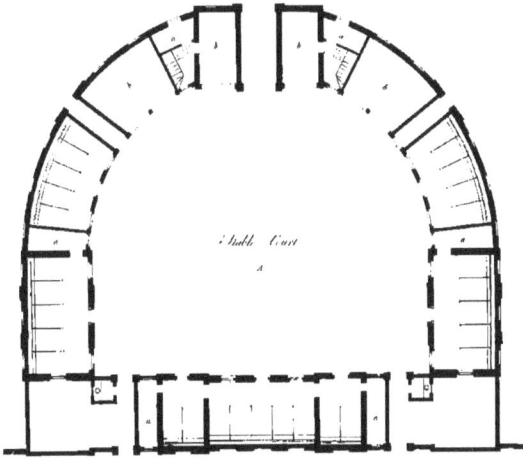

*Stable Court*

*Garden front*

Published Jan.<sup></sup> 1. 1825 by J. and J. Taylor N.º 59 High Holborn London

# · C H I L L I N G T O N ·

## · THE · SEAT · OF · THOMAS · GIFFARD · ESQ · NEAR · WOLVERHAMPTON · IN · STAFFORDSHIRE ·

The dark teints fhew the old parts.
The light teints fhew the improvements.
This houfe is built with bricks, and intended to be ftuccoed; the plinths, window-dreffings, cornices, baluftrades, and the whole of the portico are of ftone; the principal floor is clevated about two feet, and all the bafement ftory is arched.

## · P L A T E · XII ·

### · THE · PLAN · OF · THE · PRINCIPAL · FLOOR · WITH · THE · ALTERATIONS · AND · ADDITIONS ·

A portico of the Ionic order leads to the veftibule, which is decorated with columns and a vaulted ceiling; the veftibule communicates with the faloon; on the right-hand of the entrance is the eating-room, and beyond it is the library; on the left of the entrance is the withdrawing-room, which communicates with the breakfaft-room; the great ftair-cafe is next the billiard-room, and beyond it is a chamber, dreffing-room and common ftair-cafe; and all the rooms have feparate communications. An arcade leads to the fervants hall. The kitchen and offices are alfo connected with the houfe by an arcade, and the houfekeeper's apartment and butler's room are contiguous.

a. Saloon, originally intended for the chapel, and to have been extended as far as the dotted lines.
b. c. Chamber and dreffing-room.
d. Servants dreffing-room.
All the other rooms, &c. are explained in the plan.

## · P L A T E · XIII ·

### · THE · PLAN · OF · THE · CHAMBER · STORY ·

This ftory contains nine bed-chambers, c, d, f, k, o, q, y, v, t, and fix dreffing-rooms, a, e, i, p, x, u.
b. Common ftair-cafe.
g. Great ftair-cafe.
h. Corridor. m. Continuation of corridor. L L. Sky-lights. l. Sky-light in faloon; the fide windows are intended to remove the objections made to the ufe of fky-lights in rooms.
r. Stair-cafe to the new attics.
s. Water-clofet.
w. Paffage.
There is a mezzanine between the houfekeeper's apartments on the ground-floor and the rooms t, u, s, containing a chamber, dreffing-room and water-clofet.

## · P L A T E · XIV ·

### · THE · PERSPECTIVE · VIEW · OF · THE · ENTRANCE · FRONT · AS · EXECVTED ·

## · P L A T E · XV ·

### · THE · ENTRANCE · FRONT · AS · PROPOSED ·

## · P L A T E · XVI ·

### · THE · SECTION · OF · THE · GREAT · ROOM · OR · SALOON · AS · PROPOSED ·

The plan of the principal floor fhews the variations.

## · P L A T E · XVII ·

### · THE · PLAN · AND · ELEVATION · OF · THE · INTENDED · BRIDGE ·

Chillington is greatly indebted to the late Mr. Brown for one of the fineft pieces of water in England; it was the intention of the prefent poffeffor to have had another bridge built over it, according to this defign.

d

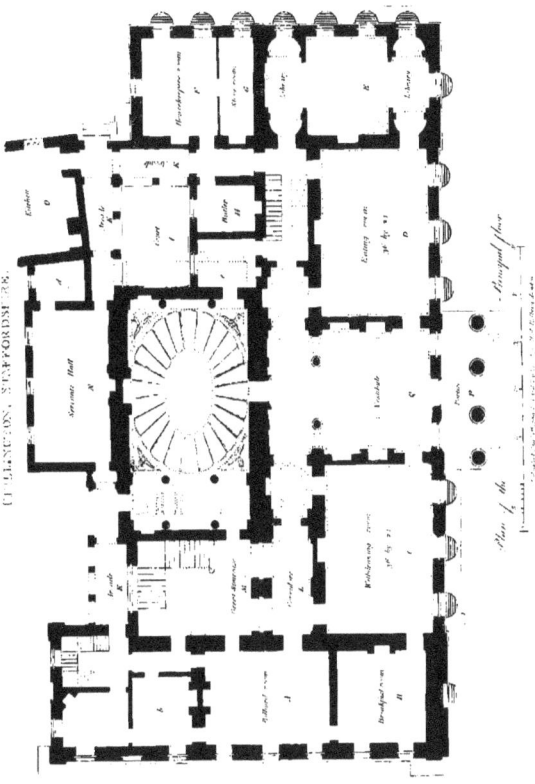

TEDDESLEY, STAFFORDSHIRE.

Plan of the Principal Floor

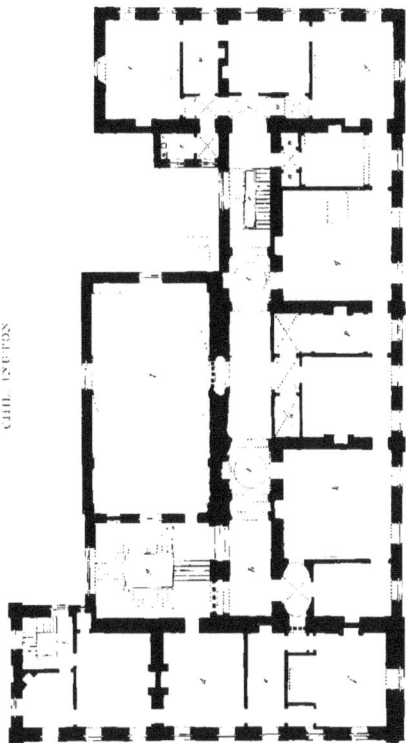

AULA REGIA.

Plan of the ● ● ● Chamber Story
Pensile

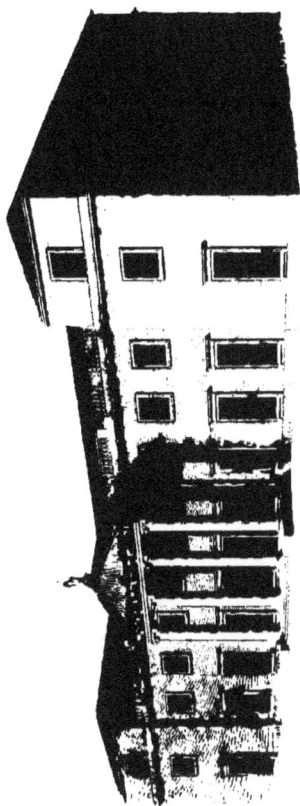

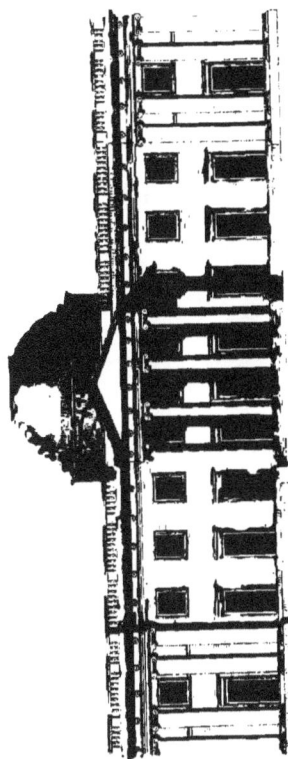

CHELSEA COLLEGE.

Eastern front  as projected.

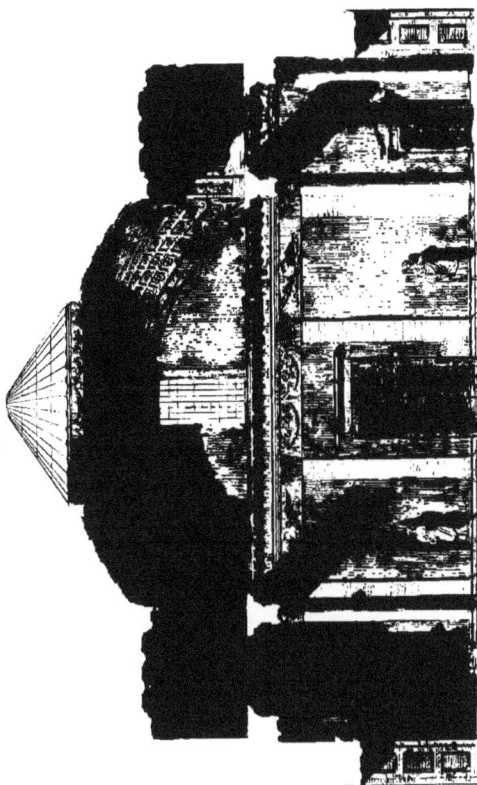

CHALMETTE

Section of the Grand Marais

*Elevation*

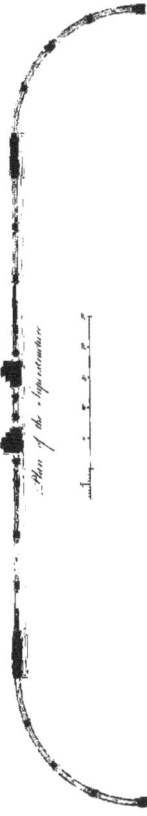

*Plan of the upper structure*

# · T E N D R I N G · H A L L ·

## · THE · SEAT · OF · SIR · JOSHVA · ROWLEY · BART · NEAR · · STOKE · IN · SVFFOLK ·

This houfe is pleafantly fituated in a park, commanding a variety of pleafing objects; the fronts are of white bricks, the fteps, planths, fafcias, cornices, and the whole of the portico, are of Portland ftone.

### · P L A T E · XVIII ·

#### · THE · PLAN · OF · THE · PRINCIPAL · STORY · AND · THE · ENTRANCE · FRONT ·

A portico of the Doric order conducts to the hall, which is finifhed with a vaulted ceiling; oppofite the entrance is the door into the great ftair-cafe; and in the fame line the approach to the withdrawing-room.— The eating-room is on the left fide of the hall, and the billiard-room is between the eating-room and the withdrawing-room; on the right of the hall is a chamber and two dreffing-rooms, and beyond them is the library; the great ftair-cafe is in the center of the houfe, and with the common ftair-cafe makes the com- munication with every room feparate and diftinct.
a. Veftibule, feventeen feet fix inches by thirteen feet nine inches.
b. Common ftair-cafe.
c. d. Dreffing-rooms.

### · P L A T E · XIX ·

#### · THE · PLAN · OF · THE · BASEMENT · STORY · AND · THE · SECTION · FROM · · NORTH · TO · SOVTH ·

The whole of the bafement ftory is vaulted.

a. Servants dreffing-room.
b. Room to clean fhoes and knives.
c. Butler's working-room.
d. Plate clofet.
e. Common ftair-cafe.
f. Lobby.
g. Scullery.

### · P L A T E · XX ·

#### · THE · PLAN · OF · THE · CHAMBER · FLOOR · AND · THE · SECTION · FROM · · EAST · TO · WEST ·

a. a. a. The young ladies apartments.
b. b. Clofets.
c. Water-clofet.
Between thefe rooms and the principal floor is a mezzanine, containing two chambers, two dreffing-rooms, water-clofet, and other conveniences.
d. Lady's chamber.
e. Powdering-clofet.

### · P L A T E · XXI ·

#### · THE · PLAN · OF · THE · STABLE · BVILDINGS · AND · ELEVATION · OF · THE · · ENTRANCE · FRONT ·

a. Stable for fick and lame horfes.
b. Stair-cafe to hay-lofts and grooms rooms.
c. Harnef-room.
d. Old tower; a room in it for preparing warm mefhes, &c.
e. Stair-cafe, &c.
f. Stable for ftrangers coming to the houfe on bufinefs.
h. h. Double coach-houfes.
i. Entrance.

### · P L A T E · XXII ·

#### · THE · PLAN · AND · ELEVATION · OF · THE · LODGES ·

Plate 16.

TENDRING HALL, SUFFOLK.

*Entrance front.*

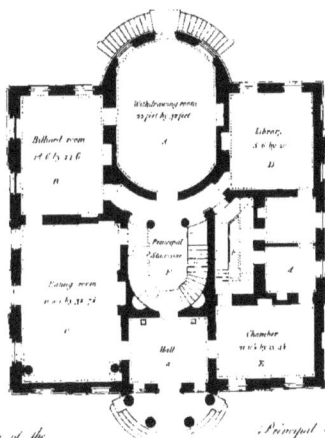

*Plan of the* *Principal Story.*

Published according to Act by I. & J. Taylor, N.º 56, High Holborn, London.

Plate III

TENDRING HALL

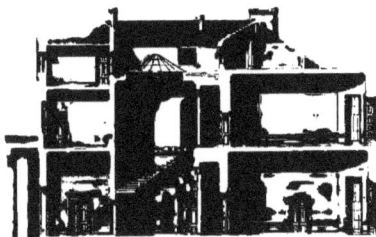

Section from North to South.

Plan of the Basement Story.

Published Jan.y 1.st 1795, by I. Taylor, No.t 59 High Holborn, London.

*Section from East to West*

*Plan of the Chamber floor*

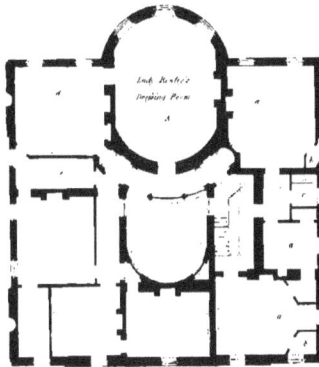

_Stable Court_

_Plan_

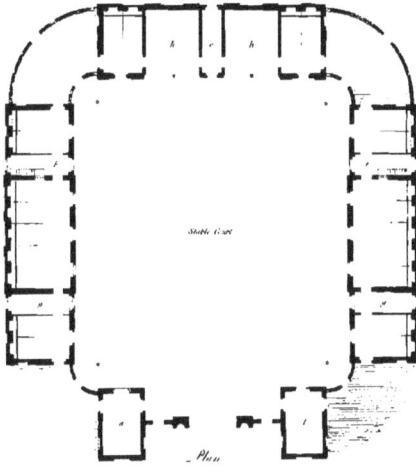

_Elevation of the Entrance front_

*Elevation next the Road*

*Plan of the Bridge*

# · L A N G L E Y · P A R K ·

## · THE . SEAT · OF · SIR · THOMAS · BEAVCHAMP · PROCTOR · · BAR^T · NEAR · NORWICH ·

THESE lodges form the entrance into the park, and are built with white bricks; the pedeſtals for the iron-gates, the columns, the entablature, and all the other ornaments, are of Portland ſtone.

## · P L A T E · XXIII ·

· THE · PLAN · AND · ELEVATION · OF · THE · LODGES ·

## · P L A T E · XXIV ·

· THE · ELEVATION · OF · A · DESIGN · MADE · BEFORE · THE · SITVATION · · WAS · DETERMINED ·

f

Plate 21

LANGLEY PARK, NORFOLK.

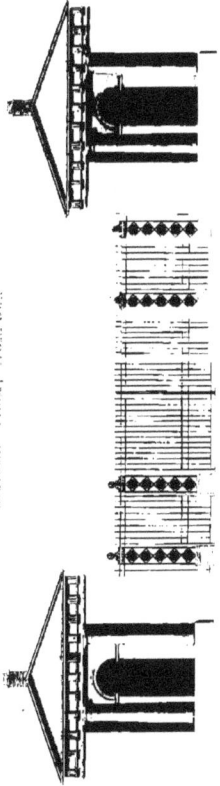

Elevation next the Road

Plan of the Lodges

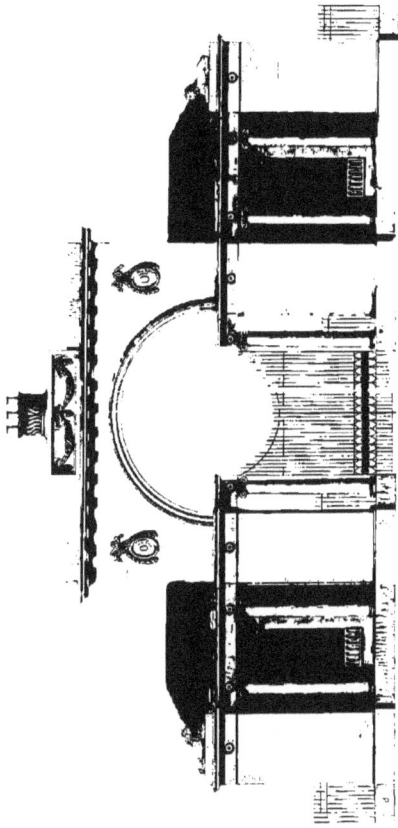

# · R Y S T O N · H A L L ·

## · THE · SEAT · OF · EDWARD · ROGER · PRATT · ESQ · NEAR · DOWNHAM · IN · NORFOLK ·

The dark teints fhew the old buildings, and

The light teints fhew the improvements.

THE principal ftory is confiderably elevated ; and the fronts are intended to be roughcafted.

## · P L A T E    XXV ·

### · THE · PLAN · OF · THE · PRINCIPAL · STORY ·

A FLIGHT of ftone fteps leads to the veftibule, and directly fronting is the door into the withdrawing-room ; on the right of the veftibule is the library, opening into the eating-room, which communicates with the withdrawing-room ; on the left of the veftibule is a dreffing-room, chamber, and ladies dreffing-room, or breakfaft-room ; the great ftair-cafe is of ftone, and is fituated between the laft-mentioned room and the withdrawing-room ; the common ftair-cafe communicates immediatly with the eating-room, veftibule, &c.

   a. Dreffing-room.
   b. Arched lobby.
   c. Great ftair-cafe.
   d. Balcony.
   f. f. f. Clofets.
   g. g. g. Servants lodging-rooms.

## · P L A T E · XXVI ·

### · THE · PLAN · OF · THE · BASEMENT · STORY ·

FROM the difference in the levels of the ground in the entrance and lawn fronts, the whole of the bafement ftory is nearly upon the level of the ground in the front of the lawn ; the great ftair-cafe is continued in this ftory.

   a. Clofets.
   b. Mr. Pratt's powdering-room.
   c. Strong clofet.
   d. Paffage of communication from the laundry offices to the kitchen offices, &c.
   e. Houfemaid's clofet.
   f. Water-clofet.
   g. Houfekeeper's ftore-room.
   h. Recefs in houfekeeper's room.

## · P L A T E · XXVII ·

### · ELEVATION · OF · THE · ENTRANCE · FRONT · OF · THE · HOUSE · AND · OFFICES ·

g

RYSTON HALL, NORFOLK.

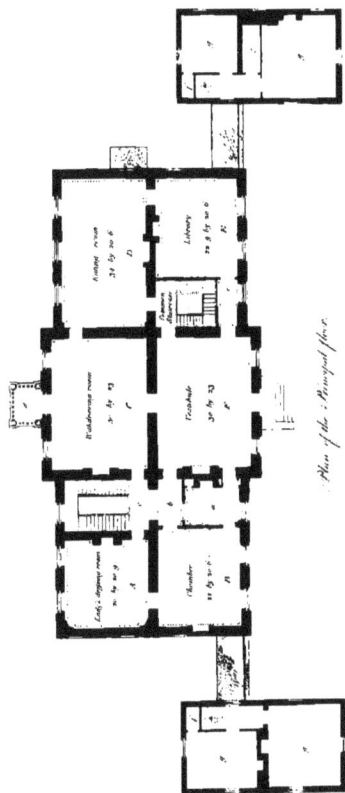

Antichamber

Eating room

Library

Common Entrance

Withdrawing room

Vestibule

Lady's dressing room

Chamber

Plan of the Principal Floor.

RESTON HALL

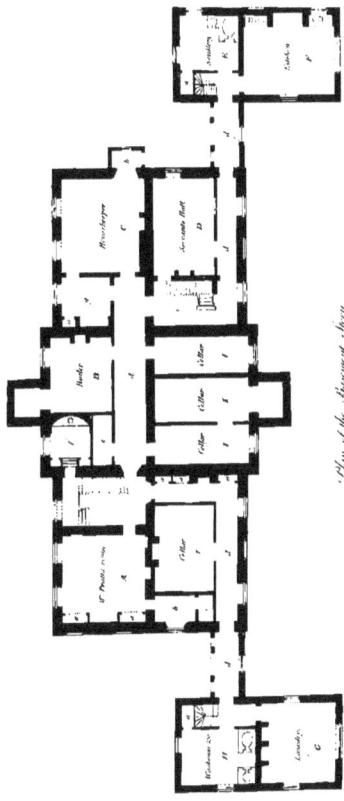

Plan of the Basement Story

Entrance d'Arsal

SKELTON CASTLE, YORKSHIRE.

*Entrance front*

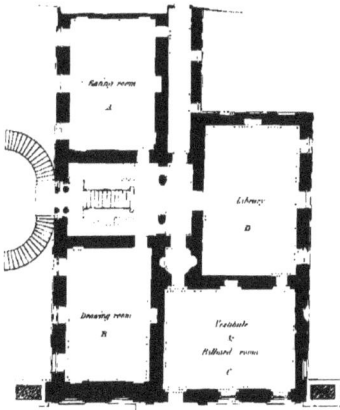

*Plan of the* &c. *Principal Floor.*

# · S K E L T O N · C A S T L E ·

## · THE · SEAT · OF · JOHN · WHARTON · ESQ · NEAR ·
## · GISBOROVGH · IN · YORKSHIRE ·

The dark teints fhew the old parts.
The light teints fhew the improvements.

THREE defigns were made for the alterations and improvements of Skelton-Caftle; the plan and elevation Nᵉ 2, are fettled to be carried into execution.

## · P L A T E · XXVIII ·

### · THE · PLAN · AND · ELEVATION · OF · DESIGN · Nᵒ · 1 ·

Veftibule and billiard-room, thirty-three feet by twenty-four feet fix.
Library, twenty-four feet by thirty-fix feet.

## · P L A T E · XXIX ·

### · THE · PLAN · AND · ELEVATION · OF · DESIGN · Nᵒ · 2 ·

Veftibule, thirty feet by twenty-two feet fix.
Breakfaft-room, twenty-three feet by twenty-two feet.
Library, forty-fix feet fix by twenty-five feet.
a. a. Receffes in great flair-cafe.
b. The great flair-cafe.

## · P L A T E · XXX ·

### · THE · PLAN · AND · ELEVATION · OF · DESIGN · Nᵒ · 3 ·

Veftibule, thirty feet by twenty-two feet fix.
Library, forty-feven feet by twenty-four feet.
c. Recefs in veftibule, decorated with columns, niches, &c.
a. Clofet for great coats, fticks, &c.
b. Great flair-cafe.
d. Common ftair-cafe.
e. Clofet for wood, &c.
f. g. Dreffing-rooms to the ftate-chamber.
State-chamber, twenty-two feet fquare.

Entrance front

Plan of the principal floor

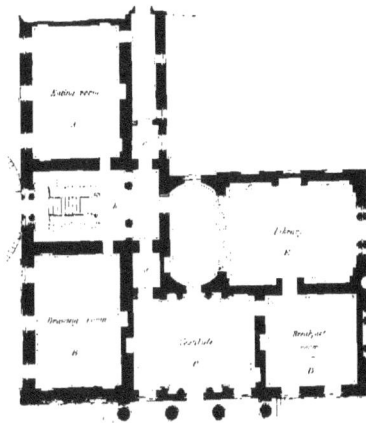

Plate 30

SKELTON CASTLE.

*Entrance front.*

*Plan of the Principal floor*

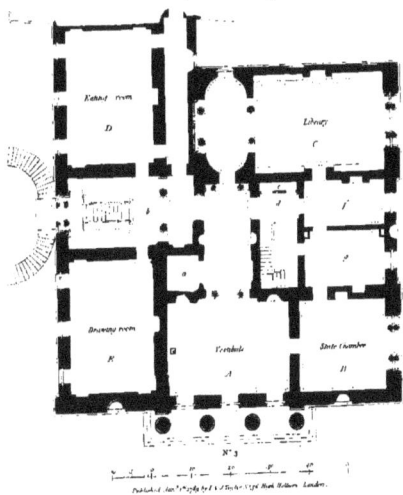

No. 3

Published Jan.'y 1 1789 by I & J Taylor Nº 56 High Holborn London.

# · M V L G R A V E · H A L L ·

## ·THE · SEAT · OF · THE · RIGHT · HON·ᴸᴱ · LORD · MVLGRAVE ·

## · NEAR · WHITBY · IN · YORKSHIRE ·

MULGRAVE-HALL is fituated on an eminence, within a fmall diftance of the fea; and commands different views of Whitby-Abbey, the remains of a caftle, and of a fine romantic country. The houfe is built entirely of ftone, and the principal floor is confiderably raifed.

## · P L A T E · XXXI ·

### · THE · PLAN · OF · THE · PRINCIPAL · STORY ·

A SMALL portico enclofes the fteps, and removes the inconvenience ufually attending the having the firft floor confiderably above the level of the ground. In the veftibule oppofite the front door is the entrance into the gallery, which communicates on the left with the withdrawing-room, on the right with the eating-room; the library adjoins the withdrawing-room; to the right of the hall is the eating-room, and on the left the beft and common ftair-cafes.

a. Lobby to library, withdrawing-room, and great ftair-cafe.
b. Great ftair-cafe.
c. Paffage.
d. Portico and fteps.
e. Anti-room to eating-room, for the fervants to attend in, &c.
f. Stair-cafe.
g. Butler's fleeping-room.
h. Paffage to offices.
i. Open court to keep the fmells from the offices out of the houfe; and alfo a way to let pipes, &c. into the wine-cellar.
k. Bakehoufe. The fmall ftair-cafe leads to the men-fervants lodging-room over the kitchen, and to the cook's chamber over the bakehoufe, which two rooms have no communication with any other part of the houfe.
l. Kitchen.
m. Houfekeeper's ftore-room.
n. Mangle-room to laundry.
o. Wet and dry larders; beyond them is a dairy.
p. Scullery, which communicates with the kitchen-court.

## · P L A T E · XXXII ·

### · THE · PLAN · OF · THE · BASEMENT · STORY ·

a. Water-clofet.
b. Paffage, containing preffes for wardrobes, &c.
c. Lobby.
d. Strong-room.
e. Wriung clofet.
f. Warm bath.
g. g. Stair-cafes.
h. h. Paffage to the offices.
i. Room for fervants to drefs.
k. Stair-cafe to kitchen offices, &c.
l. Madeira-cellar.
m. For letting down cafks, &c.
n. Coal.
o. o. o. Earth.

## · P L A T E · XXXIII ·

### · THE · ELEVATIONS ·

i

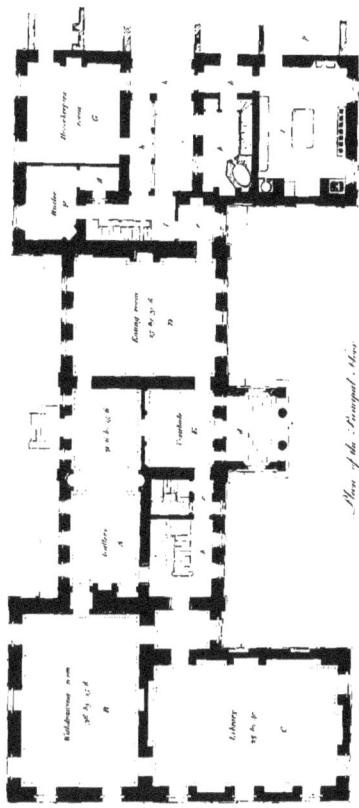

MUSGRAVE HALL, YORKSHIRE.

Plan of the Principal Floor.

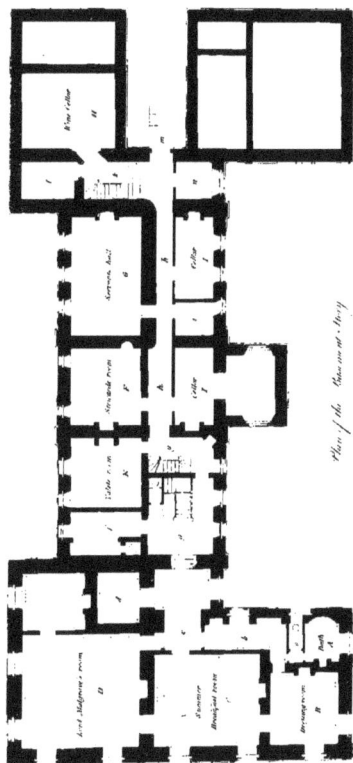

Plan of the Basement Story

Wine Cellar
Servants hall
Cellar
Cellar
House Keepers room
Servants room
Cellar
Lord Melgrave's room
Scullery
Breakfast room
Drawing room
Bath

South front No 2

South front No 1

Published Janᵗ 1ˢᵗ 1804 by I. Taylor No 59 High Holborn London.

# · B V R N · H A L L ·

## · THE · SEAT · OF · GEORGE · SMITH · ESQ · IN · THE ·
## · COVNTY · OF · DVRHAM ·

THE fituation intended for this houfe is uncommonly beautiful; the eating-room had been built fome time, and it was defigned to have completed the plan, had not the owner, in the interim, purchafed Piercefield in Monmouthfhire, which occafioned the defign to be laid entirely afide.

## ·P L A T E · XXXIV·

### · THE · PLAN · OF · THE · PRINCIPAL · STORY ·

a. Balcony.
b. Lobby.
c. Common ftair-cafe.
d. Gentleman's drelling-room.

## ·P L A T E · XXXV·

### · THE · ELEVATION · OF · THE · ENTRANCE · FRONT ·

## ·P L A T E · XXXVI·

### · THE · PLAN · OF · THE · OFFICES · AND · ELEVATIONS · OF · THE ·
### · SAME · AS · EXECVTED ·

THE whole of this building is of ftone.

a, a, a, a, a, a. Cow-houfes.
b. Arched recefs.
c. Pens for calves.
d. Bull.

k

Withdrawing room

Chamber

Library

Entrance

Eating room

Billiard room

Vestibule

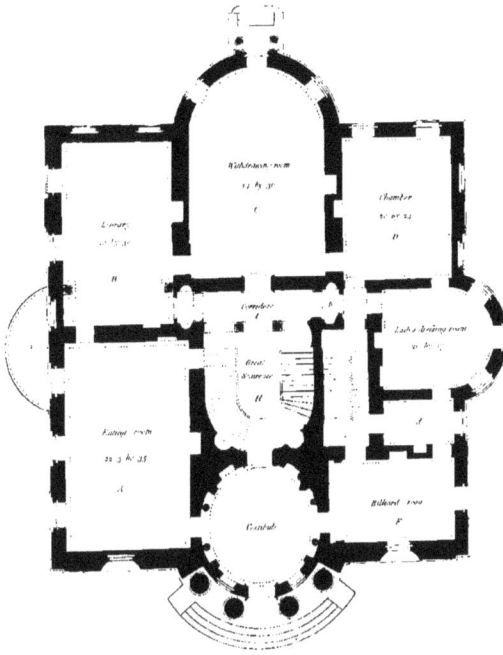

Plan of the Principal floor

Cottage front.

*Elevation of the back front.*

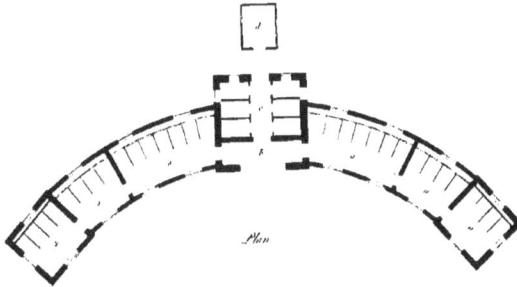

*Plan*

*Elevation of the Entrance front.*

# · VILLA · NEAR · HOCKERIL ·

## · BELONGING · TO · RALPH · WINTER · ESQ ·

This houfe and offices are built with ftudwork; the ground-floor is confiderably elevated; the fronts are roughcafted, and the roof is covered with flates.

### · P L A T E · XXXVII ·

#### · THE · PLAN · OF · THE · GROVND · FLOOR · AND · THE · ELEVATION · OF · · THE · ENTRANCE · FRONT ·

Drawing-room, eighteen feet by fifteen feet.
Eating-room (exclufive of recefs), fifteen feet by twenty-two feet fix.
Kitchen, eighteen feet by feventeen feet.
Wafh-houfe and bakehoufe, thirteen feet by eleven feet fix.
a. Porch.
b. Clofet.
c. Stair-cafe.
d. Larder.
e. Communication between houfe and offices, and common ftair-cafe, to the rooms over the kitchen and wafh-houfe, &c.
f. Porch to offices.
g. Scullery.
h. Meal-room.
i. Oven.

# · OVLTON · NEAR · LOWESTOFFE · IN · · SVFFOLK ·

## · THE · SEAT · OF · NATHANIEL · RIX · ESQ ·

This houfe is built with bricks; the outfide walls are roughcafted; the principal floor is raifed about four feet from the ground, and to prevent the inconvenience attending a large number of external fteps, part of them are made in the paffage.

### · P L A T E · XXXVIII ·

#### · CONTAINS · THE · PLAN · OF · THE · GROVND · FLOOR · AND · THE · · ELEVATION · OF · THE · ENTRANCE · FRONT ·

a. Entrance.
b. Store-room.
c. Larder.
d. Mr. Rix's room.
e. Stair-cafe.
The eating-room has a vaulted ceiling.
The kitchen is level with the ground, and over it are two lodging-rooms for fervants.
The other offices are in part under the houfe.

1

*Elevation of the Entrance front*

*Plan of the Ground floor*

Published Jan.ʳ 1.ˢᵗ 1789 by Isaac Taylor N.ᵒ 56 High Holborn London.

*Elevation of the Entrance front.*

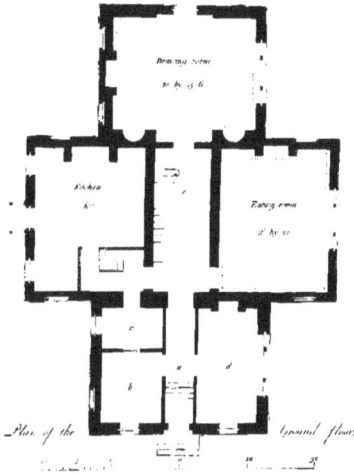

*Plan of the Ground floor.*

·P L A T E · XXXIX ·

·D E S I G N · N° · 1 ·

· THE · PLAN · AND · ELEVATION · OF · A · VILLA · FOR · THE · HON<sup>BLE</sup> ·
· WILBRAHAM · TOLLEMACHE · INTENDED · TO · BE · BVILT · AT ·
· MOTTRAM · IN · CHESHIRE ·

a. Water-clofet.
b. Great ftair-cafe.
c. Clofet.
d. Ditto.
e. Common ftair-cafe.
f. Paffage to kitchen and offices.
g. Cook's clofet.
h. Scullery.
i. Larder.

·P L A T E · XL ·

·D E S I G N · N° · 2 ·

THE · PLAN · AND · ELEVATION · OF · ANOTHER · DESIGN · FOR · THE ·
· SAME · SITVATION .

a. Coals for kitchen,
b. Arcade between kitchen offices and houfe.
c. Beft ftair-cafe.
d. Common ftair-cafe.
e. Recefs in eating-room.
f. Arcade between houfe and laundry-office.

*Elevation of the Entrance front*

*Plan of the Principal floor N°3*

*Published for ... by ... John High Holborn London*

Elevation of the interior front

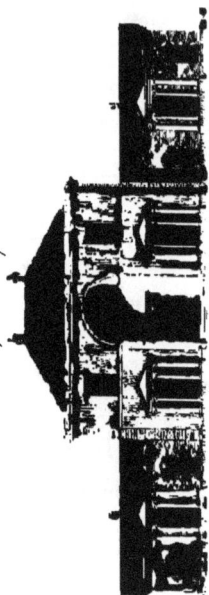

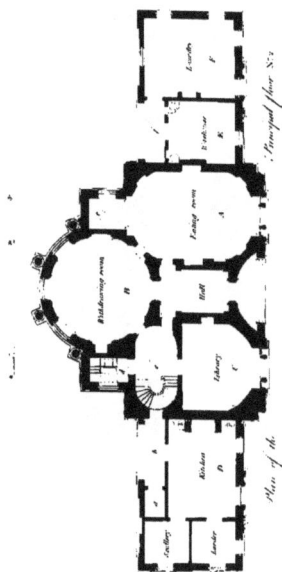

Plan of the Principal floor Nº 1

Drawing room  B
Hall
Library  C
Eating room  A
D. roof  K
Larder  F
Kitchen  D
Larder
Scullery

# · THE · PARSONAGE · AT · SAXLINGHAM ·

## · BELONGING · TO ·
## · THE · REV<sup>RD</sup> · ARCHDEACON · GOOCH ·

This houfe is fronted with white bricks, and the principal ftory is raifed about two feet. One of the wings was intended to have contained the kitchen offices and houfekeeper's room, and the other the coach-houfe and ftables; but in the execution the offices were entirely changed.

The dark teints fhew the houfe.
The light teints fhew the offices, &c.

## · P L A T E · XLI ·

### · PLAN · OF · THE · PRINCIPAL · STORY · AND · ELEVATION · OF · THE ·
### · ENTRANCE · FRONT ·

This houfe contains, on the principal ftory, a veftibule, eating-room, drawing-room, ftudy, and two ftair-cafes; the fituations of which are fhewn in the plan.

a. Pantry.
b. Scullery.
c. c. Paffages from the offices to the houfe.
d. Houfekeeper's room.
e. Court.
f. China-clofet.
g. Court.
h. Neceffary.
i. A light clofet for the convenience of the eating-room.

## · P L A T E · XLII ·

### · CONTAINS · THE · PLAN · OF · THE · PRINCIPAL · STORY · AND · THE ·
### · ELEVATION · OF · THE · ENTRANCE · FRONT · OF · THE · FIRST ·
### · DESIGN · PROPOSED ·

a. Stair-cafe.
b. Hall.
c. Portico.

n

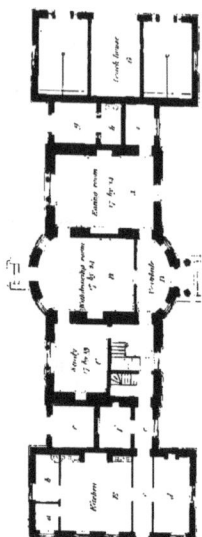

THE PARSONAGE AT SANDELFORD, NEAR YORK.

*Elevation as proposed*

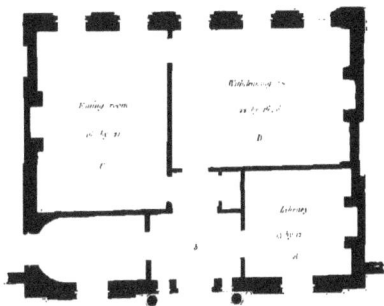

*Plan of the                    Ground flr*

# · BLACK · FRIARS · BRIDGE · NORWICH ·

THE old bridge, confisting of three arches, being too much ruined to admit of reparation, the Corporation refolved to have a new one of Portland ftone; and as great weights would be conftantly paffing over, it was neceffary to have the new bridge as flat as poffible, without injuring the navigation.

## · P L A T E · XLIII ·

### · THE · PLAN · AND · ELEVATION ·

The chord line of the arch is forty-two feet.

THE foundations of the abutments are piled and planked. The vouffoirs of the arch have their joints worked perfectly fmooth, and are fet dry in milled lead, and in the middle of each joint of each vouffoir are inferted two cubes of caft iron of three pounds weight, let equally into each ftone, and channels are funk from the tails of the vouffoirs to the cavities for the iron joggles, and the whole of the cavities and channels are run full with lead; the fuperftructure is finifhed with iron-railing.

THE whole expence of pulling down the old bridge and building the new one was one thoufand two hundred and ninety pounds.

THE fteps next to St. George's, Bridge-Street, are not executed, as the houfes are clofe to the bridge.

o

BLACK FRIARS BRIDGE, NORWICH.

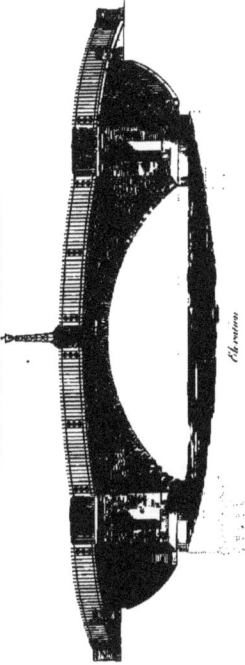

*Elevation*

*Plan of the Superstructure*

Published Jan. 1, 1818, by Jos. Taylor, N°. 9, High Holborn, London.

# · THE · DAIRY · AT · HAMMELS ·

## · THE · SEAT · OF · PHILIP · YORKE · ESQ ·
## · NEAR · PVCKERIDGE · IN · HERTFORDSHIRE ·

This building is placed near the houfe, and furrounded with large trees; the fronts are rough-cafted, and the roof is covered with reeds; the pillars are the trunks of trees, with the bark on, decorated with woodbines and creepers.

## · PLATE · XLIV ·

### · THE · PLAN · AND · ELEVATION ·

The ceiling of the loggia is arched; the dairy has alfo a vaulted ceiling, enriched with large funk pannels, filled with rofes, and other ornaments in ftucco; the tables for the milk are of marble.

Loggia, nine feet fix inches by five feet fix inches.

a. Dairy, fourteen feet fix inches by feventeen feet.

Strawberry-room, twelve feet fix inches by twelve feet fix inches; the walls are varnifhed and decorated, and the windows are of ftained-glafs in lead-work.

# · EARSHAM · NEAR · BVNGAY · IN · SVFFOLK ·

## · THE · SEAT · OF · WILLIAM · WINDHAM · ESQ ·

This edifice terminates a lawn; was originally intended for a greenhoufe, and completed for that purpofe, but has been fince converted into a mufic-room; the front is of Portland-ftone, enriched with columns, niches, and other ornaments.

## · PLATE · XLV ·

### · CONTAINS · THE · LONGITVDINAL · SECTION · OF · THE · BVILDING ·
### · WITH · THE · ALTERATIONS ·

The ceiling is highly finifhed with ftucco ornaments in compartments, as are alfo the circular ends; the walls are ftuccoed and decorated with paintings in chiaro ofcuro and other enrichments. The chimney-piece is of white marble, and the floor is paved; it being the wifh of the poffeffor to have the building as elegant as poffible.

P

*Entrance front*

*Plan*

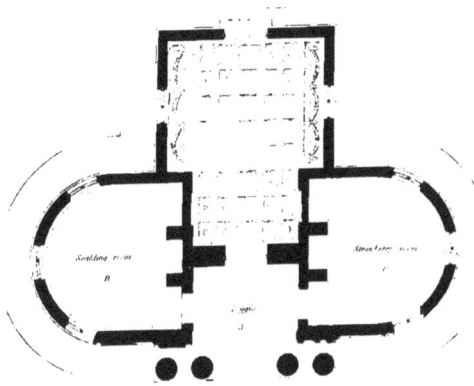

Plate 47

EARSHAM, NEAR BUNGAY.

Longitudinal Section

Published Jan.y 1.st 1833. by J. & J. Taylor, High Holborn, London.

# · A · BVILDING · PROPOSED · AS · A · MVSEVM · FOR · THE · ⸱ DILETTANTI · SOCIETY ·

This defign was to apply two unfinifhed houfes in Hereford-Street, adjoining Camelford-Houfe, to the accommodation of the Dilettanti Society; it was the intention of the noble owner, the Right Honorable THOMAS LORD CAMELFORD, to have prefented them to the Society for the public advantage, but, on confideration, the members thought their finances unequal to fuch an eftablifhment, the idea was therefore relinquifhed.

## · P L A T E · XLVI ·

· THIS · PLATE · CONTAINS · THE · PLANS · OF · THE · TWO · PRINCIPAL · · STORIES ·

## · P L A T E · XLVII ·

· AN · INTERIOR · VIEW · OF · PART · OF · THE · MVSEVM ·

DILETTANTI.

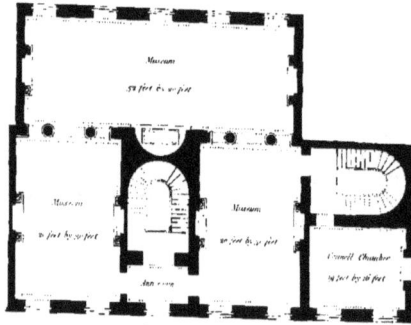

Plan of the Principal Story.

Plan of the Ground floor.

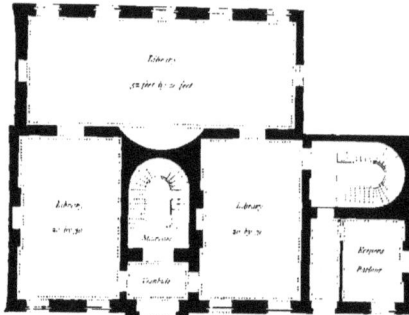

Maryland Street.

Published June 1st 18.. by J. & J. Taylor, No. 56 High Holborn, London.

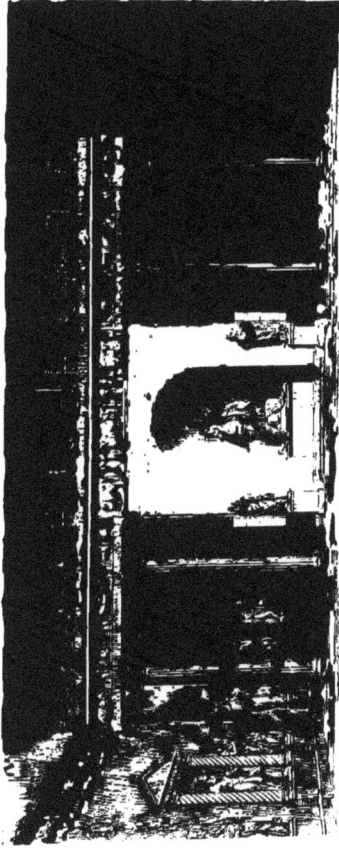

CHE TEANTL.

TO THE RIGHT HON.ᵇˡᵉ T. GRENVILLE, &c. &c. &c.
this plate, in grateful acknowledgment of his friendship, is dedicated
by his much obliged & faithful servant